the guide to owning a

Sugar Glider

Sue Fox

T.F.H. Publications, Inc.
One TFH Plaza
Third and Union Avenues
Neptune City, NJ 07753

This book has been published with the intent to provide accurate and authoritative information in regard to the subject matter within. While every precaution has been taken in preparation of this book, the publisher and author assume no responsibility for errors or omissions. Neither is any liability assumed for damages resulting from the use of the information herein.

ISBN 0-7938-2158-4

www.tfh.com

Contents

Sugar gliders are popular, exotic pets that are native to Australia, New Guinea, and the surrounding areas.

Introduction

Exotic pets such as the sugar glider appeal to an animal lover's desire for something unusual. People who are attracted to and share their home with animals enjoy the sense of novelty and discovery that comes from owning an exotic pet. It can be both rewarding and challenging to care for one.

Many people acquire a sugar glider because they like its cute face and striking looks. For the right person, the adorable, big-eyed sugar glider makes an excellent pet. The glider's soft, silky fur is pleasant to touch. Sugar gliders bond to their owners and tame gliders ride comfortably on a shoulder or in a pocket. However, owning an exotic pet is sometimes trendy and you should be sure you will still be interested in your pet after the fad has passed.

It is always best to read about the kind of care an animal requires and what type of pet it makes before you purchase it, not afterward.

This book introduces you to sugar gliders, their dietary requirements, unique habits and activity levels, and describes how to care for them. Hopefully after reading this book, you will know whether you will be able to provide the appropriate care for a sugar glider and whether or not sugar gliders are the right pets for you.

DOMESTICATED VERSUS EXOTIC

A familiar domesticated animal is the dog. Of the hundreds of dog breeds, some breeds, such as the Dachshund, differ so radically from their wild ancestors you might never guess they were related. Fancy pet rats have been domesticated for more than a hundred years. They are also different from their wild ancestors,

Many people are attracted to sugar gliders because of their cute expressions, big eyes, and sense of novelty.

being more docile and much less aggressive. In the 70 years that chinchillas have been kept in captivity, they too, have changed. Wild chinchillas can jump eight feet high, but domesticated chinchillas can only jump about three feet.

Animals that have historically been domesticated were easy to feed and bred readily in captivity. They did not have unpredictable dispositions, did not bite people, and did not become dangerous as they grew older. Many animals that people might potentially find useful and desirable cannot be domesticated. Such animals cannot be kept in captivity for various reasons, such as their dietary requirements, mating habits, dangerous dispositions, growth rate, or tendency to

Sugar gliders are generally docile enough to be pets for people of all ages. All sugar gliders should be hand-tamed to make them easier to handle.

panic when kept in enclosures. Because these animals are difficult to keep in captivity, people cannot even begin the process of domestication with them.

In comparison, an exotic animal is not domesticated. The animal might have been imported from its native habitat or bred in captivity, but it is not naturally tame or friendly with people. Such an animal still has strong natural instincts and might have unpredictable changes in temperament. For example, it might react defensively when protecting its offspring or territory (such as its cage). Generations of selective breeding are required to alter traits such as temperament so that the species might be more naturally friendly and less fearful with people. Unlike a domesticated animal, you cannot be assured an exotic pet will behave in a predictable way.

Sugar gliders are exotic animals. Because they are generally docile, they have the potential to make good pets. But while sugar gliders are suc-cessfully kept as pets, remember that it does not come naturally to them to want human company. No matter how gentle your sugar glider might seem and no matter how much time you spend socializing it, your glider is a tamed wild animal. It will take time and patience to make a wild animal into a friendly pet that is not afraid of people.

Many gliders that were tame at one time, or with one person, will revert to their wild nature if their owner loses interest in them. If you lack the dedication to keep your pets tame, they will be more difficult to interact with when you do have time, which might make you even more likely to neglect them.

Tame sugar gliders possess certain traits that make them desirable pets such as their cuddliness, ability to bond with people, and relative ease of care in captivity. The good thing about sugar gliders is that they have been kept as pets long enough for people to know what to expect in terms of their pet qualities.

About Sugar Gliders

DESCRIPTION

Sugar gliders measure about 12 inches from their nose to the tip of their tail, and at least half of this length is their tail. Gliders use their tail for balance when climbing and gliding, but their tail is not prehensile and they do not hang from it. Gliders have five toes on their front feet. Their hind feet have four toes and a large

A sugar glider's tail is used for balance when climbing and gliding. The tail generally makes up half of the sugar glider's length.

opposable toe. The two toes adjacent to the thumb are fused together so they look like one toe with two nails. The glider uses this "syndactylous" toe to groom itself. It scratches itself with these toes, and then tidily cleans the toes in its mouth. It continues to groom and clean itself in this fashion until it is satisfied. Gliders also groom each other with their mouths. Their ears can rotate individually and seem to constantly move when gliders are awake. These insectivores have a total of 40 teeth, including six incisors in their upper jaw and four in the lower jaw.

SCIENTIFIC NAMES

Scientific names are the currency by which scientists from around the world refer to the same species of animal. Scientists and serious hobbyists use scientific names because an animal can have several different common names. An animal's common name can vary by country and can even be the same as another animal's common name. For example, the sugar glider is also called a honey glider, and the people who live in the glider's native countries might call it Tupai or Sege-sege.

Taxonomic categories are scientists' method for ordering all living things. The taxonomic groups are the following: phylum, class, order, family, genus, and species name. A species is the basic unit of classification. Genera (singular is genus) are groups of species that share a common ances-

The type of sugar glider offered for sale in pet stores goes by the scientific name *Petaurus breviceps*. *Petaurus* means "rope dancer" and refers to the glider's ability to glide through high trees.

The key to raising a sugar glider is to learn all you can about the species before you bring one home. Be sure you can provide for all your new pet's needs.

tor. Genera are grouped into families, families into orders, and so forth. Scientific classification can seem unnecessary and boring for a pet owner. However, this system is useful to you. Should you research more information on sugar gliders, you might find your pets referred to only by their scientific name.

MARSUPIALS

While most small mammals sold as pets are rodents, sugar gliders are marsupials. Sugar gliders, kangaroos, wallabies, and koalas are all marsupials. Marsupials are commonly known as a type of mammal who raise their young in an external pouch, or mar-

supium. Most marsupials have a very short gestation period. Females give birth to tiny, immature young that have a much longer period of development within their pouch. Once they are born, the babies crawl into the mother's pouch and attach themselves to her nipples. Then they continue to grow and develop by drinking their mother's milk. Young marsupials in a mother's pouch are called "joeys."

Marsupials are a type, technically an order, of mammal. Approximately 250 species of marsupials are found in the world, with most occurring in the southern hemisphere. In comparison to marsupials, mammals whose young grow and develop via a placenta are called placental or eutherian mammals.

The sugar glider offered for sale in pet stores (*Petaurus breviceps*) is native to the forested areas of northern and eastern Australia, New Guinea, and surrounding islands. The sugar glider was introduced into Tasmania in 1835 and has since spread over the island. Some scientists think that people also introduced sugar gliders into other parts of their current distribution. The sugar glider is common in Australia, although habitat loss has caused local declines.

Three other species belong to the genus *Petaurus*, which means "rope dancer" and refers to the animal's quick movements on tiny twigs high up in the forest canopy. The animals in this genus are sometimes collectively referred to as lesser gliding possums. The species name "*breviceps*" means "short head."

IN THE WILD

Sugar gliders are arboreal, which means they live in trees. Folds of skin that extend from their paws to their ankles, called a *patagium* or gliding membrane, enable them to glide from tree to tree. They can glide down more than 150 feet between trees. Their long furry tail helps them steer as they glide from branch to branch. At the end of each flight, the glider pulls up and makes a four-paw landing, using its sharp nails to grip the tree. Sugar gliders cannot glide up—they always glide with a slight downward direction. When it's time to launch again, the glider climbs higher, so that each time it descends between trees.

Sugar gliders are social animals that live together in groups. Variable group sizes have been noted in the wild. A typical group contains up to seven adult males and females and their offspring. Members of a group are thought to be related to one another. Groups are mutually exclusive, territorial, and agonistic toward each other. In captivity, established groups are known to attack newly introduced individuals. One or two dominant, usually older males defend the group's territory and mate with the females. Young gliders disperse from the group when they are between 10 to 12 months old. During cold weather, gliders huddle together to conserve energy and may even simultaneously enter daily torpor when winter food is scarce.

Sugar gliders are nocturnal and their large round eyes are adapted for seeing in the dark. During the day, they sleep in leaf-lined nests inside tree hollows. The sugar glider collects leaves while hanging by its hind feet. It moves the leaves from its front feet to its hind feet, and in turn, to its tail, which then coils around the nest material. Because the tail cannot then

Sugar gliders are arboreal. In the wild, they spend most of their lives in trees and can glide more than 150 feet.

Your sugar glider will need a large cage with plenty of climbing surfaces and places to explore.

be used for gliding, the glider brings the leaves back to its home by running along branches. Scientists who study wild sugar gliders have noted a strong, distinctive odor that permeates their nests.

Sugar gliders communicate with each other chemically by scents produced from skin glands. The dominant male glider marks other members of the group by rubbing his scent glands on them. Males often greet other gliders by rubbing their head, which has a prominent gland, beneath another glider's chin, and they mark objects by passing their bellies back and forth on it. Gliders also scent mark with their urine as a means of communicating. Gliders also communicate by vocalizations, including an alarm call that resembles the yapping of a small dog and a high-pitched call when angry or frightened.

Before You Buy a Sugar Glider

Many people are attracted to the sugar gliders' adorable looks. However, sugar gliders are more demanding to care for than many other small pets. Carefully consider the following points before you buy a sugar glider. If you think you cannot meet these requirements, you might want to consider keeping a different type of pet.

Most sugar glider owners hand-tame their pets. Your new glider may need time to adjust to being handled. Be patient and persistent, and in time, you will see results.

EXPENSE

Although sugar gliders have increased in both popularity and numbers, they are still expensive pets to own. Because sugar gliders cannot be kept alone, you must budget for at least two animals. In addition, the large cage they require is expensive. Besides these one-time expenses, you will have food costs and the recommended veterinary care.

TAMING

Unlike domestic pets, it takes time, patience, and often, considerable effort to make a sugar glider into a companionable pet. Not all sugar gliders are nice. Many individuals will become wonderful pets, but others might always be frightened or mean, no matter how much effort you spend. Because of this possibility, a sugar glider might not be the right pet for you. However, many pet owners enjoy the rewarding experience of taming a sugar glider so that it is a well-socialized, bonded pet. When taming a glider, be aware that it might nip and bite. Some will bite hard enough to draw blood and it can be difficult not to swat the glider away when it bites you. Sugar gliders can be complex, and they might be friendly toward one person, but not another. This can present problems if you ever need to find a new home for your pet.

TIME

Sugar gliders require daily care, and not just during taming and bonding sessions. Their food takes time to prepare. Fresh food must be offered each day and removed the following morning. They are not as fastidious as many small pets, and their cage might need to be tidied each day. Furthermore, it is highly recommended that sugar gliders be allowed to exercise outside their cage each evening.

LIFESPAN

Sugar gliders are a long-term responsibility. These hardy little pets live much longer than similarly sized animals such as gerbils and hamsters. In the wild, sugar gliders can live for four to five years; however, sugar gliders in captivity can live up to 14 years.

NOCTURNAL

Sugar gliders are nocturnal. Many gliders do not wake up until nine or ten o'clock at night, and some might not wake up until midnight. Their nocturnal schedule complements many owners who are away at work or school during the day. However, for people who are not night owls, they can be difficult pets with which to interact. Their instinctive nocturnal behavior cannot be turned off. Gliders do not like normal daylight or bright lights. Trying to keep them awake during the day will not work, as they will just fall asleep.

Before you get your heart set on owning a sugar glider, make sure that they are legal in your state. Laws regarding the keeping of "exotic" pets can vary from state to state. Also be sure that there is a vet in your local area who can treat your pet if he becomes ill or injured.

Doing so can also stress them, which will lead to health problems.

However, some pet owners do get their pets to wake up slightly earlier by altering the light cycle so that the room containing the gliders' cage is dark earlier than normal late in the day, and light earlier than normal early in the morning. To be successful this process must be consistently adhered to and requires such modifications as darkened windows and timers.

Young children in particular are unlikely to stay up late enough to enjoy

A sugar glider's large eyes are adapted for a nocturnal lifestyle. Your sugar glider should be allowed to sleep during the day and may be irritable if you wake him up early.

pet gliders. Although they are often called pocket pets (which refers to the small animals traditionally considered children's pets), sugar gliders are not recommended as pets for young children unless the gliders are the entire family's pets.

LEGAL STATUS

Sugar gliders are considered exotic animals and are not legal to keep as pets in all states. For example, they are illegal in the state of California. Although national legislation governs the keeping of animals, individual states still have their own laws, and these laws vary from state to state and even from county to county. States can ban personal ownership of an animal species for one or more reasons, or they might require special permits to keep an animal. Generally, if pet stores in your area can sell sugar gliders, they are legal. To be certain of the gliders' legal status, check with your state's Department of Wildlife.

FINDING A VETERINARIAN

Sugar gliders have been available as pets for less than ten years. Because they are still novel in some areas, not all veterinarians can be expected to be familiar with gliders. Many veterinarians have no experience in exotic animals, but some vets are interested in them and willing to care for them. Find a qualified veterinarian before you purchase your sugar glider. Sugar gliders are expensive pets and it is foolish to not have a veterinarian available to treat them. Ask for recommendations from the pet shop or breeder where you purchased your pets.

Selecting Your Sugar Glider

WHERE TO BUY A SUGAR GLIDER

Sugar gliders can be purchased from a variety of sources. The most reliable places are either pet stores or local breeders. It is best if you buy your pets close to where you live because long-distance shipping can be stressful for gliders. Pet fairs and expositions are less desirable venues since the person from whom you buy your glider might not be available for recourse should the pet turn out to be ill or of nasty temperament.

Buy your pets from a reputable pet store or breeder, either of whom should give you a health guarantee

You can purchase sugar gliders from pet stores or breeders. Be sure to examine the sugar glider before you bring him home. He should look lively and alert and show no sign of illness.

for 24 or 48 hours. A federal license is required to sell baby gliders so make sure the seller has an exotic pet license with the United States Department of Agriculture (USDA).

Buy your glider from someone who is knowledgeable about these animals. Be wary of people who handle gliders with gloves or those who cannot satisfactorily answer you questions. Do not buy your pets from a crowded or dirty cage, or from a cage where the sugar gliders are improperly housed. For example, sugar gliders should not be sleeping on the ground. Their cage should have a nest box and branches.

Gliders kept under improper conditions are more likely to develop stress-related health problems. Although the gliders will be sleepy, be sure to handle the animals in which you are interested. Before you take your new pets home, be sure to find out what they have been eating.

PRICE

As previously stated, sugar gliders are expensive. Baby gliders that have been handled from an early age cost more than those that have not been socialized. Untamed gliders cost less than tame ones. Because of their high price, sugar gliders are unlikely to be an impulse buy. Nonetheless, do not let an inexpensive animal lure you into a purchase you might regret. Be leery of bargains. There is usually a good reason that a glider is cheap; it might bite, it might fight with other gliders, it might not like people, it might be sick, or it might be too old

Sugar gliders are social animals. They live in medium-sized groups in the wild. A single sugar glider kept as a pet will become lonely and stressed. Buy more than one and they will keep each other company when you are not home.

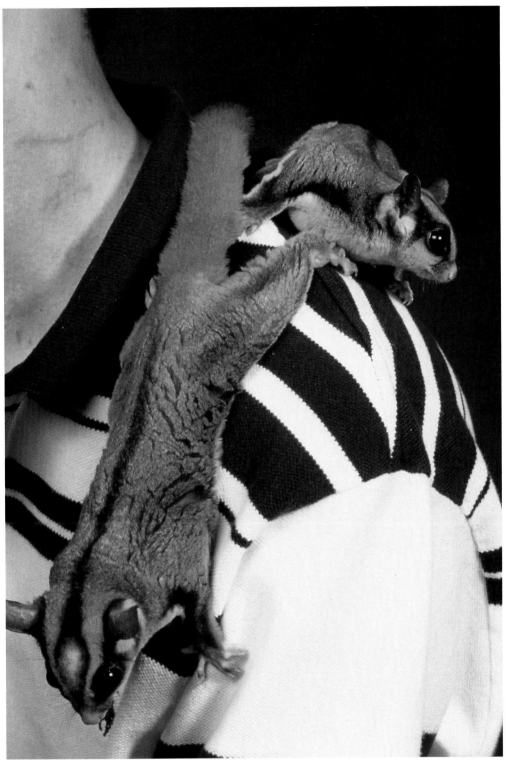

A sugar glider of either sex will make a great pet, and both sexes are equally affectionate to their owner. If you purchase a pair of gliders and do not want them to breed, have the male neutered.

to breed. Sugar gliders are not bred in fancy colors, but if new color types are developed, you can expect that they will cost more than the standard color.

HOW MANY?

Sugar gliders are social animals that live in family groups. They should not be kept as single pets. They need other gliders for their well being. A glider kept by itself can fail to thrive; it might lose its appetite and become depressed and lethargic. Single animals are not more likely to bond to their owner. You should buy your gliders at the same time so they can best adapt to each other. Gliders groom each other, play with one another, and sleep in their nest box together.

Same sex groups get along fine. If housing more than one male together, the males must either be littermates, introduced at a young age, or neutered. If you keep a male and a female together, they will produce young unless the male is neutered. It can be difficult to introduce additional gliders to an established group at a later time.

MALES VERSUS FEMALES

Male and female sugar gliders are sexually dimorphic, which means they differ in size. Adult male gliders tend to be heavier than adult females (115 to 160 grams compared to 95 to 135 grams). However, their body and tail length is often similar.

It is relatively easy to determine a sugar glider's gender, even when an individual is only a few months old. The male has a bifid penis, which is normally retracted and therefore not visible. However, his scrotum is easily observed as a small sac like bulge on his belly, often where pet owners think a belly button should be. In comparison, the female glider has an abdominal pouch opening on her belly.

As they mature, gliders develop secondary sexual characteristics. By six months of age, male gliders typically develop a bare patch of skin on their head, which is a scent gland. Both males and females have scent glands located on their chest and alongside their cloaca. A cloaca is a common passageway for the products of an animal's digestive, urinary, and reproductive systems.

Both male and female sugar gliders make excellent pets. No significant differences in life span have been noted between male and female gliders. The most significant difference in "pet qualities" is that males tend to be more pungent smelling than females. However, neutering a male can reduce the musky, fruit odor. Females can be spayed so that they cannot breed, but the operation is more complex, and therefore more risky, than neutering a male. Because

Signs of a healthy sugar glider include: bright, clear eyes, a bushy tail, smooth, dense coat, and a good weight. This sugar glider appears to be in good physical shape and looks lively.

females are not as pungent as males, you would be unlikely to notice any difference in a female's smell even if she was spayed.

CHOOSING A SUGAR GLIDER

The sugar gliders sold in the pet trade did not originate from Australia because Australia banned exports of its native wildlife more than 40 years ago. Imported gliders are usually collected from Indonesia. Wild-caught gliders are still imported and typically have brown, rather than gray, fur. It is thought that their fur is stained from plants and their nests because they eventually turn gray.

Purchase pet gliders that were born in captivity, sometimes referred to as domestic-bred gliders, not wild-caught gliders. Imported, wild-caught gliders have several disadvantages such as an unknown age, stress from shipping, and adjusting to a new diet and environment. They are also unlikely to ever tolerate handling. Imported individuals are best suited for established breeders who are looking to expand their genetic lines.

The best gliders to buy are those that have received regular handling from a young age so that they are well socialized with people. Gliders with the best pet potential have been handled as babies shortly after they emerged from their mother's pouch and opened their eyes. The breeder will have handled the gliders each day until they are ready to be weaned.

A sugar glider should be at least six weeks out of the pouch before being weaned. This means that your choice

Take your time when choosing a sugar glider for a pet. Ask questions about the animal's health, behavior, and diet, and find out if it has been hand-tamed.

Any sugar glider you adopt should be handled daily from a young age. A sugar glider that has not been socialized may be frightened of humans. Hand-taming will take time and patience.

should be almost four months old. To avoid potential problems, do not purchase a glider younger than this age.

Finding a recently weaned sugar glider that has been handled daily can sometimes be difficult. You might have to get on a waiting list, but the wait is well worth it. A well-socialized sugar glider is more likely to bond to you, be friendly, and adjust to its new home. Sugar gliders less than ten weeks out of the pouch are usually easier to tame than older individuals. It will take more effort to tame a frightened, unsocialized youngster compared to one that has been handled at an early age.

If the secondary sexual characteristics of the male are evident, such as the scent gland on his head, then you know he must be at least six months old. Adult gliders can be tamed, but it can be difficult, and sometimes impossible. Untamed adults will usually learn to take treats from your hand and occasionally they will become tame enough to pet and hold.

Be sure the sugar glider you choose to adopt has clear, bright eyes, no nasal discharge, good eating habits, and has a bushy tail. Bright-eyed and bushy-tailed applies here, because gliders with sparsely furred tails are probably too young to be weaned. A glider should feel solid, not frail. Its fur should be dense and soft, with no obvious signs of thinning.

Sometimes people find they do not have enough time for their pet gliders. These older animals are not a bad choice as long as they are healthy and friendly. Gliders that have been repeatedly passed from home to home probably will be stressed and be more difficult pets.

SELECTING YOUR SUGAR GLIDER

Housing Your Sugar Glider

Sugar gliders are active animals. In the wild, the male's average home range encompasses 1.2 acres, or about 53,000 square feet. Although they are similar in size to gerbils, they require a much larger cage in comparison to their body size. Even if you plan on letting your sugar gliders out of their cage each evening to play, their cage must still be large enough to allow them to run, leap, and climb. Their cage must be spacious enough to also include a nest box, tree branches, and an exercise wheel without being crowded.

THE CAGE

Choose a cage constructed of wire mesh. The space between the wire bars should not be greater than a half inch by one inch. If there are only vertical wire bars, make sure the space between the bars does not exceed half an inch. Do not house your pets in a glass or plastic aquarium. Such enclosures are typically not large enough, do not provide sufficient ventilation, and if not set up correctly, can force the tree-dwelling gliders to sleep on the ground. Furthermore, gliders seem to enjoy climbing up a cage's wire bars and hanging upside down from the wire by their feet to survey their domain.

Naturally, the size of the enclosure you need depends on the number of sugar gliders you plan to keep in a single cage. The more animals you have, the larger the cage must be. An aviary that measures six feet high, six feet long, and four feet wide is the ideal sized cage for a colony of sugar gliders. Such a cage can comfortably house up to seven adults. Of course, an aviary's large size makes it impractical for most pet owners. If you do

Sugar gliders need a cage large enough to allow them to glide from branch to branch. A tall, sturdy cage made of wire is ideal for sugar gliders.

Your sugar glider will love to play and climb all over the wire cage bars. The cage should have a secure lock so your glider doesn't escape.

not have room in your house for an aviary, choose a cage that measures at least three to four feet high, two to three feet long, and two feet wide. Such a cage is suitable for a pair of sugar gliders.

Although still uncommon, some manufacturers are now making cages specifically for sugar gliders. Otherwise, you should look in the bird department of a pet store to select your gliders' cage. Cages sold for finches and cockatiels are usually suitable for sugar gliders. Although cages made for parrots are commodious, the openings between the wire bars are typically too far apart to be suitable for sugar gliders (and parrot cages can be very expensive).

Your pets' cage should have a solid bottom tray constructed of either durable plastic or metal. Two styles of wire cages are commonly available, with either a slide-out tray or a tray that detaches from the wire enclosure. Either type is suitable for your gliders. Some cages have a wire screen floor that allows an animal's droppings to fall through to the tray below. Because gliders spend little time on the ground of their cage, being in direct contact with their droppings and bedding is not a significant concern.

Look for features that make a cage more desirable, such as lightweight, wheels on very large cages, door(s) large enough for you to easily reach your hand inside, and a top that opens for easy access to the upper half of the cage. Check that there are no sharp edges on which your pets or you could get cut.

Gliders are curious animals and will look for ways to escape their cage. They can squeeze out small openings so cage doors must close securely. If you have any doubts, use an additional latch to secure the cage door. Some clever gliders can figure out how to open the doors on birdcages that slide up. Such doors need an additional latch.

Many pet owners make their own cages. As long as you construct a sturdy cage, this is a reasonable option. You can attach the correct size cage wire mesh to either a wood or metal frame of the desired dimensions. Be sure to file any sharp edges down. Be aware that some sugar gliders will chew on the wood. Attaching the wire mesh to the inside of the wood frame can prevent them from doing so.

BEDDING

A variety of small animal bedding, including wood shavings (pine, aspen), crushed corncob, recycled paper, and wood pulp products can be used on the floor of your sugar gliders' home. Place a thin layer of the bedding material on the bottom of your gliders' cage. By absorbing liquid waste (and water from the occasional leaking bottle), bedding keeps animals warm, clean, and dry. Because sugar gliders are arboreal they do not spend much time in direct contact with the bedding in their cage. Bedding is not necessary to provide warmth for your pets; its main purpose is to absorb and provide a medium for collecting waste. You can even line the bottom of their cage with plain paper if you choose. Since most newspaper inks are now vegetable-based, not petroleum-

The cage you choose should be large enough to give all your gliders room to play and climb. Don't clutter it with too many accessories.

Even though sugar gliders are arboreal, you still need to provide them with bedding material. A thin layer of pine shavings or recycled newspaper bedding will help keep the cage warm, clean, and dry.

based, they are not detrimental to small animals. Check with your newspaper's printer to be certain.

Do not use sawdust, which is too dusty for your pets. Dusty bedding can irritate a pet's respiratory system or aggravate an existing respiratory ailment. The potential effect of dust on a pet often depends on whether or not the animal is housed directly on the bedding (called contact bedding) or whether it is housed above the bedding in a wire-frame cage (called non-contact bedding). Because gliders are unlikely to burrow and dig in their bedding, they are not likely to stir up fine dust particles. In general, paper pulp and recycled paper products tend to be lower in

dust compared to wood shavings. Cedar shavings have been implicated as both causing and aggravating respiratory problems in small animals and should not generally be considered a good choice for your gliders.

Keep your sugar glider's bedding clean. The harsh ammonia odor that develops in your pets' cage is uncomfortable for both you and your pet. Ammonia is a severe irritant and is detrimental to the health of pets. It affects the mucous membranes of the eyes and respiratory tract and it leaves animals open to opportunistic infections. Gliders housed on dirty, moist bedding are most susceptible to these effects.

While some bedding contains odor-masking agents, such as chlorophyll in shavings, the development of new, innovative bedding products has been spurred in part by the quest to control or eliminate odor. Because these materials do not just mask odor, they promote a healthier environment compared with traditional bedding. Many of the newer products cost more, but they are also scientifically engineered to control ammonia formation, are super absorbent, and need to be changed less often than other materials. However, these beddings will not eliminate the natural glider smell that is present in your pets' cage.

ACCESSORIES

Your sugar gliders need a nesting box for security and to sleep in during the day. Various types of suitable nest boxes are sold at pet stores, including plastic hamster houses and wooden bird nest boxes. Some pet owners prefer plastic houses, which are easier to keep clean than wooden boxes that absorb urine and other smells. A properly designed model should open from the top in addition to the round entrance hole for the animals. This allows you to easily clean the box, and provides ready access to your pets (if necessary). You can also make your own nest box out of a

Your sugar glider should have a cozy place to sleep during the day. A hollowed out log or small box will do just fine. Line the inside with bedding to keep your glider comfy.

plastic margarine container or an empty cereal box.

For coziness and absorbency, provide bedding inside your pets' nest box. Gliders will urinate in their nest boxes, which can acquire a strong odor if they are not regularly cleaned or replaced. Securely attach the nest box near the top of the cage, but leave enough space for your gliders to comfortably perch on top of the box.

When first taming sugar gliders, it is sometimes easiest to provide a nesting pouch for them to sleep in during the day. The whole pouch can then be removed from the cage so that the gliders can be readily taken outside. However, be cautious when providing either a commercially sold nest made of cloth, such as a ferret hammock, or a homemade cloth nest. Loose threads can accidentally wrap around a glider's leg, cutting off circulation, and their sharp nails can get caught in clothing seams, causing the animal to panic and potentially tear its nail while struggling.

Your sugar gliders need branches in their cage. Branches allow them to exercise in a more natural manner. They will jump and climb on the branches, which helps to keep their nails trim, and they will chew off the bark looking for insects (and sap). Safe, suitable branches are sold in the bird section of pet stores and include manzanita, aspen, willow, and euca-lyptus. Branches of certain species of trees should not be used, as they are potentially poisonous. These include almond, apricot, black walnut, cherry, and peach trees. Arrange the branches so they provide maximum use of the space within your pets' cage, leaving enough room for the gliders to jump from branch to branch.

TOYS

Your pets will enjoy playing with and investigating a variety of toys. In general, many of the toys made for parrots can safely be used for sugar gliders. Items such as ladders, swings, chains of metal or plastic, balls, bells, and even mirrors can make your pets' cage more interesting and provide entertainment for your pets. Toys made for hamsters constructed of durable hard plastic, such as plastic crawl tubes, are also suitable for sugar gliders. Additional items to consider include hollow wooden logs sold in the reptile section of pet stores and cardboard rolls from paper towels. Wooden items might need to be replaced occasionally because gliders will chew on them and the wood will pick up odors. Gliders also enjoy gracefully sprinting along natural fiber ropes in their cage.

Many gliders enjoy playing and running on exercise wheels. Some individuals might run on a wheel for several hours each night. Sugar gliders are more likely to use an exercise

Some sugar glider owners provide their pets with hanging pouches to sleep in. The entire pouch can be removed from the cage if you want to take your glider out for some exercise or handling.

All sugar gliders need toys to play with. You can find a variety of interesting playthings at your local pet store.

wheel if they are introduced to one as youngsters. An appropriately sized wheel should measure between 9 to 12 inches. A solid floor wheel, rather than a wire screen running surface, should be used with sugar gliders. Because their long tails can accidentally get caught in the cross braces of traditional wire exercise wheels, choose a model that is specially designed for long-tailed pets. Several styles are available, including one that looks like an old-fashioned film reel with multiple entry holes for easy access by your pets. Securely mount the wheel to a cage wall in the upper half of your pets' home. Well-constructed exercise wheels are quiet. Appropriate exercise wheels can be found in some pet stores, advertised in specialty pet magazines, or found on the Internet under exotic pets.

WHERE TO KEEP THE CAGE

Sugar gliders do best at temperatures between 70 and 80 degrees Fahrenheit with normal household humidity. Gliders will become lethargic when the temperature is too cold or too hot. If the ambient temperature in your house is below 70 degrees Fahrenheit, you must find a way to increase the temperature in the room where your sugar gliders are housed. Space heaters for the room or radiant heat units for the cage are possible means to provide sufficient warmth.

Keep your gliders' cage in a draft-free location out of direct sunlight.

A large cage door not only allows you more access to your sugar gliders, but it helps make cleaning the cage easier.

HOUSING YOUR SUGAR GLIDER

Make sure to clean your sugar glider's cage at least once a week. Clean living conditions will help keep your glider healthy.

Do not place their cage near a heater, an air conditioner vent, a door, or in front of a window. Do not place your pets' cage on the floor. These arboreal animals are most comfortable when they can survey their surroundings from a high location. Place their cage on a sturdy stand or a table. If other pets have access to the room containing your gliders' cage, be certain they cannot harass them.

Keeping your pets' home in an area where you can easily enjoy them is important. Most gliders do not seem to be sensitive to noises made in busy rooms, such as the living room. The garage is not a suitable location for your pets. Garages contain toxic chemicals and automobile exhaust, are subject to extreme temperature variations, and you are likely to neglect your pets if they are kept in such a location.

CLEANING THE CAGE

A clean cage plays an important role in keeping your gliders healthy. Each morning you must remove any uneaten foods, especially moist foods that could attract fruit flies and other insects. A dirty environment will not only smell unpleasant, but it will also provide a breeding ground for potentially harmful bacteria and fungi.

Clean your pets' cage at least once a week. Replace all the bedding, even if it does not look dirty. Unlike many other types of small pets, sugar gliders do not use one corner of their cage as a bathroom area. Instead, they defecate everywhere and droppings might sometimes fall outside their cage. Sugar gliders urinate on the branches, toys, and wire bars of their cage, which can sometimes feel sticky. In between weekly cleanings, you can remove droppings and wipe down the cage bars to minimize any odors and keep your pets' home clean. Clean plastic toys by washing them in water, but wooden toys sometimes warp when washed too often. Placing a tray or towel beneath your pets' cage will help to contain any debris that falls out for easier clean up.

Your pets will scent-mark with urine more frequently if you continually remove all evidence of their urine. All animals have a smell. If you find your gliders' natural smell too strong between weekly cleanings, try using an odor control product, such as a box of baking soda near the cage (but removed when your pets are out playing) or a more sophisticated method such as ionizers.

Once a month, do a thorough cleaning. Scrape off or file any grime that might have accumulated on the wires of the cage. Replace any branches, toys, or dishes that are too dirty or damaged to clean and reuse. Disinfect the cage and the surrounding area. Pet stores sell mild cleansers that are safe for animals or use a 25 percent bleach solution. Be sure to thoroughly rinse and dry the cage before placing your pets back inside.

Always keep your gliders' water bottle clean. The inside of the bottle and the tube will feel slimy, even when they look clean. You can purchase a slender bristle brush to help clean the inside of the water bottle.

Feeding Your Sugar Glider

IN THE WILD

In the wild, sugar gliders are omnivorous. Their natural diet consists of arthropods, such as moths, beetles, caterpillars, and spiders; Acacia species gum (exuded on the trunks and branches of some Acacia species to bind damaged sites); the phloem sap from Eucalyptus species (sugar gliders use their enlarged lower incisors for chewing into the tree's bark); manna (a type of sap oozing from wounds at sites of insect damage on the leaves and branches of trees); honeydew (the exudate produced by some types of sap-sucking insects); and nectar and pollen (produced in usable amounts in larger eucalyptus flowers). Sugar gliders also eat an occasional small lizard, rodent, bird and their eggs.

A sugar glider's diet varies seasonally. During autumn and winter, plant exudates make up most of their diet, while in spring and summer they are primarily insectivorous. The glider's long fourth finger on its front feet helps it pull out insects from crevices. Insects that are attracted to sticky tree sap are also readily eaten. Scientists have found that manna, pollen, honeydew, and nectar are only minor components of a sugar glider's natural diet. Observations show that sugar gliders spend approximately 40 percent of their foraging time feeding on Acacia gum, 30 percent foraging for arthropods, and 11 percent obtaining sap from eucalyptus trees. One study that examined the feces and stomach contents of gliders indicates that almost half arthropods and half gum are actually eaten.

QUALITY DIETS

The importance of a good diet can-

Sugar gliders have dietary needs that are different from other small mammals, cats, and dogs. As of yet, the complete nutritional needs of some exotic pets, like the sugar glider, are not well known.

not be underestimated. The quality of an animal's diet can affect its health, appearance, reproductive capacity, longevity, and even temperament. The dietary requirements of many small animal pets such as rats and hamsters are well known because of their use as laboratory animals. For owners of these pets, it is a simple matter to open a bag of food and provide complete nutrition. However, the nutritional requirements of exotic pets are not always as well known.

Every species of animal has its own unique nutritional requirements. A good animal diet is formulated by nutritional experts' best assessments of current research, feeding trials, and scientific studies. In some cases, an animal's diet is modeled after a similar species whose nutritional requirements are known or the diet is based on what food items the animal will eat in captivity. In the absence of scientific studies establishing nutritional requirements, many feed manufacturers and breeders rely on two criteria to determine a diet's adequacy: fecundity and longevity. It is reasonably assumed that increases in both measurements are an indication of a diet's ability to meet the nutritional needs of the animal. However, such a method does not establish requirements for necessary nutrients, and neither can a diet designed in this fashion be considered to contain optimal amounts of essential nutrients.

DIET OF A PET GLIDER

It is the sugar gliders' misfortune that many individuals are not very discriminating in what they will eat. This predilection has led pet owners to allow them to dine on nuts, ice cream, crackers, bird seed, canned fruit cocktail, and other food items that lead to their eventual ill-health. In the wild, being an opportunistic eater is an advantage, but as pets it has worked against gliders. As a pet owner, you must keep in mind that just because your glider eats a particular food does not mean that it should, or that the food is a healthy part of its daily diet.

Your pets' diet should approximate what they would normally eat in the wild. Although sugar gliders might look like flying squirrels, which are rodents, they are very different. A sugar glider's digestive tract is simple and adapted for its insectivore/carnivore metabolism, whereas the squirrel's gastrointestinal tract is long and adapted for an herbivorous diet. Sugar gliders should not be fed nuts, seed diets sold for small animals or birds (including parrots), rodent seed sticks, canned fruit cocktail, or fruit juice instead of water.

A sugar glider is a bug's worst nightmare. Your cute big-eyed pet likes to crunch on bugs, even bugs that are as big as the glider itself. The nutritional value of invertebrates is variable. For example, termites are high in fat compared to some species of beetles. Wild animals usually spend a lot of time actively foraging for their prey. By contrast, captive animals do not need to spend time searching for food. Sugar gliders enjoy chasing, catching, and eating live foods. You can also catch live insects, such as grasshoppers, for your pets to eat.

SUGGESTED DIET

Because sugar gliders have been kept in zoos for many years, their nutritional and caloric needs have been extensively studied. However, this information has not been used to feed pet sugar gliders a suitable diet. Sugar gliders are hardy animals and have been fed foods that are convenient for pet owners, but do not typically provide an appropriate diet. Pet gliders have been fed a free-choice diet that allows them to pick out preferred items from a mix of fruits, seeds, mealworms, crickets, cottage cheese, yogurt, and other foods. Among hobbyists, it has been accepted that a sugar glider's diet should consist of 75 percent fruit and vegetable matter and 25 percent protein. These "home recipe" diets and some commercial glider foods have not undergone feeding trails and are not controlled for guaranteed analysis.

However, the best available information from Australian zookeepers, veterinarians, and naturalists has shown that a more appropriate diet

Your sugar glider should be fed a balanced diet of animal- and vegetable-based foods. Try offering a small amount of a new food item mixed in with the glider's regular food.

for pet gliders is based on a 50:50 ratio of Leadbeater's Mixture, which you must make yourself, and a commercially manufactured insectivore/carnivore diet.

LEADBEATER'S MIXTURE

150 ml warm water
150 ml honey
1 shelled hard-boiled egg
25 grm high-protein baby cereal
1 tsp vitamin/mineral supplement

Mix the warm water and honey. In a separate container blend the egg until it is homogenized. Gradually add the honey water mixture, then the vitamin powder, then and the baby cereal, blending after each addition until the mixture is smooth. Keep the mix refrigerated until served. You can also increase the amount you make at one time and freeze the additional mix. The mixture does not freeze hard so it is relatively easy to scoop some out of a container each night.

It is possible that as sugar gliders become more popular, feed manufacturers will produce a more convenient form of Leadbeater's Mixture. Until then, you must be prepared to spend time making part of your pets' diet.

Leadbeater's Mixture was originally designed by Australian scientists to feed captive Leadbeater's possum (*Gymnobelideus leadbeateri*), which are very similar to sugar gliders.

Find out what foods your sugar glider was eating before you brought him home and stick to that diet. If you must change your glider's diet, do so gradually.

Both species are similar in size and ecology. They both eat tree sap and insects, are nocturnal, and nest communally in trees. However, the possum cannot glide between trees and therefore consumes more calories per day than the sugar glider.

TREATS

Feeding treats to your gliders is an enjoyable activity and can help you to tame your pets. However, treats should not be more than 5 percent of a glider's daily food intake. Suitable treats include small pieces of cooked meats (such as chicken or beef), diced fruits (try grapes, apples, melon, pear, apples, grapes, mangos, papayas, carrots, and sweet potatoes) dusted with a multiple vitamin mineral powder, bee pollen (usually available at health food stores), crickets and various other insects (fed a good cricket diet with additional calcium, usually termed gut-loaded, and available at pet stores). Certain foods that are appropriate for your pet might only be available in certain seasons. Freezing the foods, such as grapes, can extend the length of time your pet has access to them. Nectars formulated and sold for lories (a nectar-eating parrot) can be given as an occasional treat. Provide the liquid treat in a bird water tube to prevent your pets from getting themselves messy.

Milk-based products, such as yogurt and cottage cheese, can be given in tiny amounts as a treat. However, it is the general consensus among veterinarians that these foods have the potential to cause diarrhea due to lactose intolerance. When gliders are fed the right diet, they do not need milk products for additional calcium. While small amounts of these foods are unlikely to cause any symptoms of lactose intolerance, problems can arise when pet owners begin to give their gliders too much of these foods.

Other live foods that you pet might enjoy include pinky mice, which are newborn mice without fur. However, besides being messy, they are not a good food choice for your pets. Pinkies have poor bone development and an imbalance in calcium to phosphorous ratio. Anoles, which are small lizards sold in pet stores, are a better choice and are relished by some gliders. Other foods to avoid are those with preservatives, pesticides, added sugar, artificial colors, and excessive fat.

HOW MUCH AND WHEN TO FEED

Some commercially prepared diets have suggested feeding recommendations. Use these recommendations as a starting point to determine how much to feed your gliders. Depending on the type of food, a glider should be offered no more than 20 percent of its body weight daily. (Your glider can be weighed at your veterinarian's office.) Your pet should not be able to choose what foods it wants from a smorgasbord of items. Foods should be chopped and blended together to reduce the likeli-

Sugar gliders will eat almost anything and have been known to over-indulge on snacks. Offer only healthy treats, and keep snacks to a minimum so your glider does not become overweight.

hood of your glider picking out only its favorite foods and eating an unbalanced diet.

Because gliders are nocturnal, they should be fed at night. Feed your pets around the same time each evening and they will soon learn when to expect their food. Uneaten foods that might spoil overnight should be removed after a few hours. Each morning, remove any uneaten moist food from the previous night's meal. If food is always left over, reduce the amount offered. If your pets become overweight, the quantity of food should be adjusted. Pet gliders often become overweight because they are offered too much food, food that is high in fats and simple sugars, and they do not receive enough exercise. Besides using a scale to determine if your pet is overweight, you can feel its gliding membrane, which should be thin and flexible, not rounded with fat.

When you have more than one animal in a cage, it is important that they all get enough food. Part of the social interactions among gliders involves access to food, with dominant animals getting more food or more of a preferred item. If necessary, provide more than one feeding area and space the food dishes far enough apart to ensure that all animals get enough food.

CHANGING DIETS

An improper diet is the main reason why a sugar glider will become sick. A protein and calcium deficient diet can cause a variety of serious ailments. Because the conditions caused by a nutritionally deficient diet can take a while to show up, gliders fed a poor diet might have a shorter lifespan. By the time symptoms of a diet-related disease appear, the disease has progressed to a serious state and can be more difficult to effectively treat.

Gliders that are fed a poor diet often become reluctant to eat anything except what they were previously fed. Changing their diet will require time and patience. Finicky gliders are likely to refuse any new foods. Eventually your gliders will eat their more nutritious food. They will not starve themselves. Do not abruptly change your gliders' diet as doing so can cause digestive upset. When switching your gliders to a more nutritionally sound diet, you must do so by gradually blending the new food with the old. Over a period of several weeks, increase the amount of the new diet and eventually eliminate the old foods.

WATER

Provide your gliders with fresh water using a gravity-fed water bottle available in pet stores. Bottles sold for hamsters are a suitable size. The best models are made of plastic and have a metal spout. The amount of water your gliders drink each night depends on what they are being fed. They will drink

Remove any uneaten fresh foods from the cage after a few hours so they don't spoil. This is especially important during warm weather.

less water when their diet consists of items with high moisture content.

Do not use an open dish to provide your gliders with water. They will track food and debris into the water, which will quickly become unsanitary. A water dish also means more work for you because it must be washed and refilled each day. However, if your gliders did not learn how to drink from a water bottle at their original home, you will need to temporarily provide them with water in a sturdy, non-tip dish. Place the water bottle so its spout is slightly above the dish. When your pets drink from the dish they will also brush against the spout and learn that it contains water. Eventually you can remove the

open dish when you know your pets drink from the water bottle.

FEEDING DISHES

Your pets need at least two food dishes, one for dry foods and one for moist foods. If you have numerous gliders in a single cage you might need multiple sets of dishes. More than one set of dishes is also convenient because you can prepare an evening's meal while one set is being washed in the dishwasher. Hang the dishes at least halfway up inside your pets' cage. Gliders do not like to eat on the ground. Do not place the dishes beneath the branches in the cage because your gliders might accidentally defecate or urinate in their food.

Sugar Gliders as Pets

BRINGING HOME YOUR GLIDERS

Sugar gliders rely on their sense of smell and a new cage devoid of any sugar glider aroma can be frightening. Take a handful or two of the bedding from their old home when you adopt your new gliders. Spreading the familiar-scented bedding in their new cage can help them settle into the strange environment. Plan to take your pets in for a veterinary exam to comply with the terms of any health guarantee. You can transport them in the box you used to bring them home, or in their nest box (with entrance hole blocked).

The veterinarian will give your gliders a general examination and assess them for potential problems such as parasites. If your veterinarian finds a problem, the seller should exchange the animal for another one. A knowledgeable veterinarian will also discuss proper care of your pets based on the most recent information available in veterinary journals and scientific literature.

Although you might be anxious to interact with your new pets, it is usually best not to disturb them for the first few days while they get used to their new environment. During this time you can think of names for your pets, feed them, talk quietly to them when they are awake, and watch them investigate their new home each evening. Occasionally, a new glider will not eat for the first day or two, but by the third day it should have a healthy appetite. Remember to offer food on a regular schedule and to continue feeding your pets' old diet until they have settled in to their new home. Thereafter, changes can be initiated if their current diet is inadequate.

Your sugar glider will be nervous when you first bring him home. Give him time to settle into his new cage.

TAMING

When you first bring home your new gliders, you might be surprised if they act frightened and defensive. Some gliders intimidate their owners by lunging, growling, and trying to bite. Remember, you probably bought your animals during the day when they were sleepy, and therefore less agitated. Just like people, gliders react differently when they are sleeping from when they are awake. During the day, sleeping gliders are typically cuddly and calm. But when they are awake at night they are extremely active, and gliders that are not tame are likely to be scared and refuse your overtures.

Start taming your gliders each night or early in the morning when they are still awake by offering them a food treat from your hand. Place your hand with the food treat inside your pets' cage. You must be patient. It will take time to develop your gliders' trust. A glider might take several days before it is willing to explore your hand and take the food treat. A few gliders might react to your hand by growling and trying to nip, or you might see your pets dart back and forth as they build up their courage to take the food. Do not startle them by reaching toward them. Once your pets trust you enough not to run from your hand, they will readily come to you for a food treat.

During the day, you should spend a few hours carrying your sleeping pets in a special sugar glider bonding pouch sold for this purpose. A front pocket on a regular shirt will also

Spend time holding your sugar gliders every day. It will help you bond with your new pets.

Hand-taming your sugar glider can take time, especially if your pet is not used to being held. Be patient and let your sugar glider become familiar with you.

work. Be careful not to forget that your pets are in their pouch or in your pocket so you do not accidentally hurt them. You can occasionally stroke your pets or offer appropriate treats. This method gets your pets used to your smell, voice, touch, and movements. (This process should only be used inside your house. Do not take your gliders outside.) A popular and often effective method to help your gliders get used to you is to give them one of your unwashed T-shirts in which to sleep each day.

If you consistently work with your gliders, they will become familiar with your scent and become more trusting. Try to gently rub and scratch them on their neck and head. As your gliders become more comfortable with you, they might reciprocate by grooming your hand with their mouth and paws.

You can hang your sugar gliders' pouch on the inside of their cage while you are taming them. Many gliders prefer to sleep in the pouch rather than a nest box. Pet owners often find it easier to take their sleeping pets from the cage by removing them and their pouch out of the cage. This is one of the easiest ways to remove your pets from their cage. Trying to remove untamed, grumpy gliders from their nest box can sometimes be intimidating when they utter their loud growls.

UNPREDICTABLE BEHAVIOR

Because they are not domesticated

A fully bonded sugar glider is a fun and loving pet that will want to spend time with you.

pets, sugar gliders can sometimes be unpredictable in their behavior. For example, while a sugar glider might have been sweet at the pet shop, it might become aggressive when you first bring it home. It might growl and become very protective of its nest box and cage. As a young sugar glider grows and matures, its behavior can also change so that a formerly friendly pet might nip on occasion. Some sugar gliders do bite. Some will only nip, but others might bite hard enough to break the skin and draw blood. Sometimes a glider might lick your finger, then nibble your finger, and then suddenly bite your finger.

Jerking your hand away does not prevent future incidents, nor does blowing on your glider, or tapping its nose. These responses often scare and stress your glider, triggering further antagonistic behavior. Gliders sometimes bite because they are afraid, they are defending their territory, or they do not like an odor on your hand. However, aggressive behavior has also been linked to poor health and inadequate stimulation so be sure you are properly caring for your pet.

Depending on the glider, it can take a week to several months before your pet is tame and bonded to you. Gliders that were socialized as babies and who had tame parents will bond quickest. Most gliders will eventually become tame, but some individuals will always be ornery. Like all pets, each glider will have a unique personality. Some are shy, while others are bold and curious. Some gliders will seek your attention while others will treat you indifferently as they race around a room to explore.

Not all gliders will be friendly with other people, which can be a problem if you ever have to find your pets a new home. To minimize this potential, introduce friends and family members once your gliders are comfortable with you. Gliders are small, delicate animals. Be sure young children are properly coached not to grab and squeeze a glider. If you know your pets might bite, be sure to warn visitors.

EXERCISE

To be healthy and happy, gliders need supervised exercise outside their cages each evening. Gliders play much more vigorously when they are out of their cage, racing up the curtains, gliding to a picture hung on a wall, jumping on the couch, and crawling up your pant leg. Each evening, your pets will wait for you to let them out to play. Do not expect your tame gliders to like being held when they are awake at night—this is their busy time.

It is best if you choose one room for your pets to play in and then "gliderproof" it. This means that you must make the room safe for your pets while also protecting the items in the room. Make sure windows, doors,

closets, and drawers are securely closed. Remove delicate items that might get knocked off shelves. Gliders might eat or damage houseplants. Plants should be removed from the room. Some houseplants are poisonous to animals, and you don't want your sugar glider getting sick. Block off furniture where gliders can hide and get crushed if someone accidentally sits on them.

Sugar gliders can fit into very small spaces. Furnace and air conditioner vents should be screened off to prevent access. Gliders might chew on electric cords and phone wires. If necessary, child safety plugs should be

Some sugar gliders are shy until they become accustomed to their living conditions and the people in their life. Be patient and understanding with your sugar glider if he hides when people come over to visit.

Sugar gliders need exercise time out of their cages every evening. They will enjoy gliding from your shoulder to the other objects in the room.

THE GUIDE TO OWNING A SUGAR GLIDER

used on electric outlets and cords and wires placed out of reach. Be certain that all window screens fit securely and have no tears through which your pet could crawl.

Be wary of open containers of water. Gliders are attracted to water and have been known to fall into toilets and drown. It is prudent to keep the toilet lid closed. Sugar gliders can move very fast and are noted for being extremely active. Do not take your pets outside. They do not like bright light and should they escape, you are unlikely to catch them.

While your pets are playing outside their cage, be wary of other household pets. Gliders will kill and eat mammals such as hamsters that are smaller than them, in addition to small birds and reptiles. Dogs and cats do not always intimidate gliders, but you must protect your gliders from these larger pets.

You can open your pets' cage door and let them out to begin their explorations. Or, if they must play in a different room, allow your gliders to crawl up your arm or slide your hand beneath your gliders so that they will move onto your hand. It can sometimes be difficult to return your pets to their cage when you are ready to go to sleep and they still want to play. Do not chase your pets around to try and catch them. They are very quick and will easily avoid you. Be subtle and clever. If neces-sary, try to tempt them into their cage with a treat if they will not let you catch them.

Once your gliders have bonded to you, they quickly become one of the most delightful small pets to own. Because you are their familiar territory, they will use you as their home base when they are let out to play. They will jump and glide to you in the midst of their explorations. Like birds, gliders prefer the highest part of your body. Sugar gliders have sharp nails, so wear long sleeves and pants to avoid being scratched when your pets land and climb on you. Gliders will enjoy battling a large peacock feather that you dangle in front of them or stalking a rabbit fur toy (made for cats with any plastic parts removed) that you move temptingly before them.

You cannot discipline a sugar glider. Trying to do so will rarely produce the results you want. Instead, you might ruin your pets' trust and make them fearful and distrustful of you and other people. Sugar gliders do not respond to voice commands. They do not learn the word "No." You can stop or distract them from doing a prohibited activity (such as chewing on furniture) by clapping your hands and making a loud noise, but you cannot force your glider to do something it does not want to do. It will nip or bite you or run away.

Sugar gliders are naturally inquisitive and will play with almost anything. A plastic tube makes an interesting toy.

SCENT-MARKING

Sugar gliders cannot be litter box trained. They defecate and urinate when they are outside their cage.

Both male and female gliders scent-mark with urine. This means that they might leave little drops of urine on whatever they crawl and climb, including on you. Scent-marking is a natural behavior and a means of communication among sugar gliders. This behavior cannot be changed. If you wait a half hour or so after your pets have woken up before letting them out of their cage, you reduce (but do not eliminate) the possibility that they will scent-mark outside their cage.

VOCALIZATIONS

Sugar gliders make at least four distinct sounds. The characteristic noise they make when angry has been compared to many things, including an electric pencil sharpener, the turning over of a high-pitched motor, a miniature chainsaw, a motorcycle engine, and marbles rolling around in a PVC pipe. Either way you will recognize the noise when you hear it, and the volume is quite impressive for such a small animal. This growling noise is made when the glider is angry or frightened. You are likely to hear it when you first get acquainted with your gliders and they are trying to discourage your friendly overtures. Their growls can be a prelude to biting, especially if they stand up on their back legs. Your gliders might frequently make this sound until they grow comfortable with you.

An adult's alarm call resembles the yapping of a small terrier dog. Gliders make their loud barking sound at night and the noise is loud enough to awaken you from your sleep. For some reason, they seem to give this call most often in the early morning hours shortly after midnight. Their plaintive barking can be brief or continue for several minutes. Some pet owners find their barking annoying because the call is repeated over and over at the same rhythm and volume. Pet gliders bark when one glider is out the cage and the others are locked in, sometimes when baby gliders are born, and perhaps because they want attention. A flashlight turned on nearby, but not shined directly at them, can make them stop barking.

When gliders seem excited or happy they might chatter for several minutes. You might hear this cheerful sound when your pet is perched on your shoulder. Both males and females chatter to one another when the females are receptive to mating. Their chattering might be loud or soft such as when gliders chatter to baby gliders that have left their mother's pouch. Hissing is a noise that gliders make when they greet one another or investigate a strange object. When alarmed, young gliders make a distinct nasal hissing sound.

INTRODUCING NEW GLIDERS

If you currently have a single glider and want to get a companion for your pet, you will need to cautiously introduce the gliders to each other. Although sugar gliders are social animals, new colonies usually form before the animals are seven months

Sugar gliders can be territorial to newcomers, so it's always best to start off with a pair.

old. Because gliders are territorial they will defend their territory, or cage, from unknown newcomers. This makes it difficult to introduce older gliders to one another, which is also why it is best to start off with two gliders, rather than one.

It is easiest to introduce new gliders to one another before they begin to develop secondary sexual characteristics, which usually happens around six months of age. Two gliders that are younger than six months, no matter their genders, will probably get along fine, although one will likely be dominant over the other. For a glider older than six months, it is safest to get a second glider that is similar in age to your original pet. Once males are older than six months, they will be incompatible, even if they are neutered.

While youngsters can usually be introduced into a single cage with only a little squabbling, you will need two cages for older gliders. Gliders can fight and seriously hurt one another should you place the new glider into the original glider's cage. Do not be fooled by sleepy gliders that appear peaceful during the day.

Once they wake up that evening, they will likely fight. Expect the introduction process to take a few weeks. First, place the two cages next to each other but not so close that an aggressive glider could bite the other through the cage. Once they have acclimated to one another, move the cages adjacent to one another. After they have adjusted again, you can try letting them meet in a neutral area that has not been scent-marked. Hopefully they will not attack each other. Have a towel handy to separate them should it be necessary. You may have to buy a new cage so both gliders must start with a new territory.

Zoos are often able to successfully introduce gliders into an established group by locking the established animals into a small cage within the aviary. This allows the new animals to establish themselves in the main aviary. The animals are not allowed to directly interact for several weeks. Once all the gliders are in the main aviary, there are still often fights. While this method is useful to know about, it cannot be used with your pets unless you have a very large aviary.

Sugar Glider Health Care

Prevention is the best method for keeping your pets healthy. There are several simple steps you can take to help your pets stay well. First, keep their cage and accessories clean. Do not let a dirty cage become a breeding ground for disease-causing organisms. Second, feed an appropriate

Do all you can to keep your sugar glider happy and healthy. Clean living conditions and a nutritious diet will generally prevent any health problems.

Find a veterinarian in your local area who treats exotic pets. Your sugar glider should have an annual checkup.

diet and do not offer your sugar glider treats that are high in fat or sugar. Third, make sure your gliders always have clean, fresh water available. Finally, attentive observation of your gliders will help you detect any changes in appetite or appearance that could indicate an illness.

Sugar gliders have been successfully kept in zoos for more than a hundred years. Zoos usually feature them in nocturnal houses so visitors can observe the gliders when they are active. Many of the health problems reported in pet sugar gliders, such as hind leg paralysis, have never been recorded in zoo gliders. This is probably because zookeepers have been more rigorous in the gliders' diet and husbandry.

Veterinarians are important sources of information on the care of your pets. Your veterinarian can review your gliders' care and diet and catch any problems before they become serious. An annual veterinarian exam of your gliders can provide a baseline of comparison should one of your pets become sick in the future. The exam can also alert you to any developing conditions. Diagnosis and treatment of a disease can be expensive. For example, X-rays might be required, which will necessitate that your pet be anesthetized for the procedure.

Infectious diseases, which are caused by viruses, bacteria, fungi, and protozoan, are not well known in gliders. As more people keep these pets, it is expected that such conditions will be better documented. Sugar gliders are susceptible to various types of tumors, so any strange lumps and bumps (which are obviously not babies in a female's pouch!) should be examined by a veterinarian.

Warning signs that could signal your glider is ill include weakness, paralysis of or dragging the hind legs, weight loss, lack of appetite, lethargy, diarrhea, constipation, loss of fur, and tremors. Any of these symptoms indicate a potentially serious illness and mean an immediate trip to the veterinarian is necessary. The following is a description of some of the conditions pet gliders can contract.

HIND LEG PARALYSIS

This condition is properly called nutritional osteodystrophy. It is the most common, serious disease seen in pet sugar gliders and is caused by a diet too low in calcium and vitamin D3 and too high in phosphorous. Such diets are often composed of 75 percent fruits and 25 percent meat. Most fruits and vegetables are low in calcium, but high in phosphorous, which inhibits an animal's body from effectively using the calcium. A proper diet prevents this condition from ever developing.

Symptoms such as dragging one or both hind legs, difficulty in staying upright, reluctance to move, and paralysis of the hind legs, are symptoms of this condition. If your pet displays these symptoms, he needs to be seen by a veterinarian immediately. If left untreated, the animal will die. X-rays will show brittle, porous bones that might have fractures. Osteodystrophy can be successfully treated if caught in the early stages. The illness is treated with calcium injections, cage rest, a corrected diet, and calcium and vitamin D3 supplementation.

Other diet-related conditions can include constipation and diarrhea. These symptoms indicate something is wrong, typically due to an inadequate diet. A healthy glider's droppings are similar to mouse droppings, and should be firm. When a glider is fed a diet that is high in fat, for example, an abundance of nuts, and low in fiber, in can develop constipation. A constipated glider will look sluggish and humped up as if in pain. Your vet can provide the corrective treatment for all these conditions and can recommend appropriate dietary changes.

TRAUMATIC INJURIES

Gliders that fight with one another are subject to traumatic injuries, such as bite wounds to the face and scratches on their eyes. Other injuries can occur when a glider is allowed unsupervised play outside of its cage. If such wounds look infected, you must take your pet to the veterinarian for treatment. In Australia, wildlife rehabilitation facilities frequently treat gliders for injuries due to cat predation. If you also have pet cats, be careful they cannot attack your pet gliders.

MITES

Several species of mites have been found to affect gliders, but they are not common. Symptoms of ear margin canker, which include crusting, scabs, shriveling, wrinkling, or loss of tissue, are sometimes mistaken for a mite infestation. This condition is actually related to improper diet and mites can be ruled out when your veterinarian takes a skin scraping.

TEETH

Unlike rodents, a sugar glider's incisor teeth do not continue to grow throughout its life. Therefore, these teeth do not need to be trimmed. Gliders fed an improper diet with too much sugary soft foods (for example, fruits) are prone to tartar accumulation and gum disease. Because of gum disease, a glider might develop abscessed or fractured teeth. A veterinarian can remove the tartar but it must be done under general anesthesia. Treatment with antibiotics and a diet change must follow.

BODY ODOR

Intact male gliders can smell pungent (variously described as musky, fruity, or like rotten bacon) and their urine can

Take time every day to examine your sugar glider for any injuries or signs of illness. If you suspect that your sugar glider is sick, take him to the vet.

THE GUIDE TO OWNING A SUGAR GLIDER

If your sugar glider's nails grow too long, they can get snagged on the carpet or curtains. Keep your glider's nails trimmed down.

be particularly strong smelling. These smells are unpleasant to some pet owners. In addition to odor control products such as room ionizers, neutering males will reduce the smell and their tendency to scent mark with urine. A veterinarian can neuter a male glider at around six months of age. Do not bathe your gliders with soap and water to reduce any smell. Pet stores sell products made especially to neutralize a glider's smell that you can try.

NAIL TRIMMING

A sugar glider's nails are designed to help it climb and land safely when it glides. Their nails grow quickly and are very sharp. Branches in your pets' cage will help their nails stay trim. However, a glider's long nails can get caught on clothing or a carpet, which might cause the glider to tear its claw to free itself.

A glider's long nails can painfully scratch or puncture your skin. For your pet's safety and your comfort, you might need to occasionally trim your glider's nails. (Do not use sandpaper in your pet's cage, branches, or exercise wheel to dull their nails.)

Plan to cut your glider's nails during the day when he is sleeping and calm. You can use cat clippers or regular "human" nail clippers. Two people are necessary for this procedure. Gently wrap the glider in his pouch or in a towel. If your pet begins to wiggle and squirm, wait until he falls asleep in the pouch. Then pull one foot out at a time and quickly trim the pointy tips off each nail. Do not cut into the nail's pink quick, which has sensitive nerve endings and blood vessels. Be careful not to hurt your pet, because doing so will make future attempts at cutting his

You can clip your sugar glider's nails yourself or have the veterinarian do it for you.

nails more difficult. If you are uncomfortable with clipping your glider's nails, your veterinarian can show you how.

SELF-MUTILATION

Single gliders or those that are stressed often develop pathological behaviors such as self-mutilation. These animals will bite themselves on their bodies or tails to the point that they draw blood and the tissue in the affected area begins to die. Besides their tails, gliders might mutilate their limbs, penis, and scrotum. A veterinar-ian can provide treatment so the wounds can heal, including an Elizabethan collar, which will temporarily prevent the glider from self-mutilation. Males that are self-mutilating their penises should be neutered. Disturbed gliders may also engage in repetitious behaviors, such as running in circles over and over in the same direction within their cage, or become very aggressive. Unless the underlying reasons for the behavior are addressed and corrected, such as poor diet, inadequate exercise, or a cage that is too small, the behavior will return.

Resources

American Sugar Glider Society
21418 Park Post Lane
Katy, TX 77450
Website: www.glidersociety.com
Email: asgs@glidersociety.com

Glidin' For Love Sugar Gliders
1566 Elmwood St.
Clearwater, FL 33755
Phone: (727) 443-7598
Email: Glidinforlove@aol.com

Wildlife Rescue Foundation
1265 Tyler Way
Sparks, NV 89431
Phone: (775) 284-WILD
Website: www.wildliferescue.com
Email: JPotash@WildlifeRescue.com

International Sugar Glider Association
Provides information on club membership, health care, and exotics veterinarians.
Website: www.isga.org
Email: isgaorg@hotmail.com

National Alternative Pet Association
Provides information on exotics and links to clubs and rescue organizations.
Website: www.altpet.net

Resources and links to more information on Sugar Glider breeders and organizations can be found at:

www.sugarglider.net

or

www.sugarglider.com

Index

Photo Credits

Isabelle Francais: 1, 19, 23, 48, 51, 57
Ralph Lermayer: 9, 11, 15, 18, 25, 27, 28, 45, 50, 58, 61
W. P. Mara: 30, 31, 36
John Tyson: 7, 17, 21, 29, 33, 47, 54, 55